INDUSTRY STANDARD
OF THE PEOPLE'S REPUBLIC OF CHINA

Code for Design of Energy Conservation for Railway

TB 10016-2016

Prepared by: The Third Railway Survey and Design Institute Group Corporation
Approved by: National Railway Administration
Effective date: September 1, 2016

China Railway Publishing House

Beijing 2019

图书在版编目(CIP)数据

铁路工程节能设计规范:TB 10016-2016:英文/中华人民共和国国家铁路局组织编译. —北京:中国铁道出版社,2019.1
 ISBN 978-7-113-24918-2

Ⅰ.①铁… Ⅱ.①中… Ⅲ.①铁路工程-节能设计-设计规范-英文 Ⅳ.①TU2-65

中国版本图书馆 CIP 数据核字(2018)第 200912 号

Chinese version first published in the People's Republic of China in 2016
English version first published in the People's Republic of China in 2019
by China Railway Publishing House
No. 8, You'anmen West Street, Xicheng District
Beijing, 100054
www. tdpress. com

Printed in China by Beijing Hucais Culture Communication Co., Ltd.

© 2016 by National Railway Administration of the People's Republic of China

All rights reserved. No part of this publication may be reproduced or transmitted in any form or by any means, electronic or mechanical, including photocopying, recording, or by any information storage and retrieval systems, without the prior written consent of the publisher.

This book is sold subject to the condition that it shall not, by way of trade or otherwise, be lent, resold, hired out or otherwise circulated without the publisher's prior consent in any form of binding or cover other than that in which it is published and without a similar condition including this condition being imposed on the subsequent purchaser.

ISBN 978-7-113-24918-2

Introduction to the English Version

The translation of this Code was made according to Railway Engineering and Construction Development Plan of the Year 2017 (Document GTKFH [2017] 185) issued by National Railway Administration for the purpose of promoting railway technological exchange and cooperation between China and the rest of the world.

This Code is the official English language version of TB 10016-2016. In case of discrepancies between the original Chinese version and the English translation, the Chinese version shall prevail.

Planning and Standard Research Institute of National Railway Administration is in charge of the management of the English translation of railway industry standard, and China Railway Economic and Planning Research Institute Co., Ltd. undertakes the translation work. Sichuan Lanbridge Information Technology Co., Ltd. and China Railway Engineering Consulting Group Co., Ltd. provided great support during translation and review of this English version.

Your comments are invited and should be addressed to China Railway Economic and Planning Research Institute Co., Ltd., 29B, Beifengwo Road, Haidian District, Beijing, 100038 and Planning and Standard Research Institute of National Railway Administration, Building B, No. 1 Guanglian Road, Xicheng District, Beijing, 100055.

Email: jishubiaozhunsuo@126.com

The translation was performed by Ma Liya, Zhu Haijun, Dong Suge, Chen Guocai.

The translation was reviewed by Chen Shibai, Wang Lei, Huo Baoshi, Liu Dalei, Han Lihe, Xu Weiping, Gan Jiandong.

Notice of National Railway Administration on Issuing the English Version of Four Railway Standards including *Code for Design of Environmental Protection for Railway*

Document GTKF [2019] No. 6

The English version of *Code for Design of Environmental Protection for Railway* (TB 10501-2016), *Code for Design of Energy Conservation for Railway* (TB 10016-2016), *Code for Design of Fire Prevention for Railway* (TB 10063-2016) and *Code for Design of Concrete Structures of Railway Bridge and Culvert* (TB 10092-2017) is hereby issued. In case of discrepancies between the original Chinese version and the English version, the Chinese version shall prevail.

China Railway Publishing House is authorized to publish the English version.

National Railway Administration
January 28, 2019

Notice of National Railway Administration on Issuing Railway Industry Standard (Engineering and Construction Standard Batch No. 3, 2016)

Document GTKF [2016] No. 23

Code for Design of Energy Conservation for Railway (TB 10016-2016) is hereby issued and will come into effect on September 1, 2016. *Code for Design of Energy Conservation for Railway* (TB 10016-2006) is withdrawn.

China Railway Publishing House is authorized to publish this Code.

National Railway Administration
May 26, 2016

Foreword

Based on *Code for Design of Energy Conservation for Railway* (TB 10016-2006), this Code is prepared by summarizing and drawing on the practical experience gained over recent years in design of energy conservation for railway, considering the national requirements on energy conservation, resource conservation and environmental protection and extensively soliciting opinions from all sides, as well as through review and revision.

The Code consists of nine chapters, namely General Provisions; Railway Line and Traction Power; Station and Yard; Building Construction, Heating, Ventilation and Air Conditioning; Locomotive Facilities, Rolling Stock Facilities, EMU Facilities, Machinery and Power Equipment; Electrification; Electric Power; Water Supply and Drainage; Communication, Information and Signaling.

The main revisions are as follows:

1. Addition of the requirements on implementing the energy conservation assessment report and implementing the approval review report in "General Provisions".

2. Addition of the requirements on setting pusher grade section in a concentrated way, selecting ruling grades in two directions and designing CWR for main line of new or upgraded railway, as well as revision of some previous clauses, in Chapter "Railway Line and Traction Power".

3. Summarization of the previous content according to "General Provisions", "Station" and "Depot (post), Workshop and Work Section", as well as addition of the principle that the station/depot (post) shall be constructed by stages according to the short-term demands and the long-term development with respect to the design of station/depot (post), and addition of the requirements that CWR shall be designed for receiving-departure tracks in stations of high-speed railway, intercity railway, heavy-haul railway, and mixed traffic railway with EMU operation, in Chapter "Station and Yard".

4. Addition of the requirements that thermal insulation measures shall be taken between external wall of the heating house and the ground in the severe cold zone and the cold zone, as well as specification of the requirements that electromechanical equipment such as the HVAC system shall comply with the national energy efficiency index, and specification of the requirements that monitoring and control measures for energy conservation shall be taken, in Chapter "Building Construction, Heating, Ventilation and Air Conditioning".

5. Addition of the requirements that compressed air stations shall be constructed and shared when the freight car repair yard at station is adjacent to the train inspection yard, and buildings for the wheelset tread diagnosis equipment and for the dynamic testing equipment of pantograph shall be built together, as well as revision of the content about the energy conservation design concerning locomotive, rolling stock and EMU depots (running sheds) and operation maintenance depots of large track maintenance machinery based on the actual situation of design in recent years,

in Chapter "Locomotive Facilities, Rolling Stock Facilities, EMU Facilities, Machinery and Power Equipment".

6. Revision of the original clauses according to the development of high-speed railway and inter-city railway in recent years and the change in access of power supply from public grid to the electrified railway power system in Chapter "Electrification".

7. Addition of the requirements on the use of intelligent metering devices and remote monitoring system in power consumption metering, as well as revision of the original clauses based on the proven energy conservation and emission reduction technologies and based on the energy conservation measures, in Chapter "Electric Power".

8. Revision of the original clauses according to the energy conservation and emission reduction technologies and policies such as water conservation and reuse of reclaimed water in Chapter "Water Supply and Drainage".

9. Addition of a new section titled "General Requirements", specification of the requirements that buildings for communication, information and signaling equipment of the station shall be set in a concentrated manner, and buildings for the section signaling relay station and the signaling equipment of block post and buildings for the communication base stations shall be built together, and addition of the content about energy conservation in respect of turnout control, in Chapter "Communication, Information and Signaling".

We would be grateful if anyone finding the inaccuracy or ambiguity while using this Code would inform us and address the comments to The Third Railway Survey and Design Institute Group Corporation (No. 10, Zhongshan Road, Hebei District, Tianjin, 300142) and China Railway Economic and Planning Research Institute Co., Ltd. (No. 29B, Beifengwo Road, Haidian District, Beijing, 100038) for the reference of future revisions.

The Technology and Legislation Department of National Railway Administration is responsible for the interpretation of this Code.

Prepared by:
  The Third Railway Survey and Design Institute Group Corporation

Drafted by:
  Li Xiaobing, Nie Hongwang, Li Gengsheng, Ji Huimei, Xu Lingyan, Jiang Qingping, Li Qingsheng, Song Zhiyue, Zhao Hui, Li Guiping, Yang Li, Zhang Lu, Wang Ying, Dong Zhijie, Xu Hong, Chen Weimin, Liu Yeqing, Zhou Yan, Sun Haifu, Fu Shouhua, Li Ronghua, Tao Ran, Zhu Jianzhang, Bo Haiqing, Feng Quanzai, Yang Zhenlong, Ma Jingbo, Wu Guohua, Feng Jingran and Wang Haizhong.

Reviewed by:
  Liu Yan, Tian Yang, Fu Jianbin, Han Yan, Zhao Yuexia, Chen Fangrong, Wang Zhonghe, Zheng Qingsong, Long Weimin, Liu Xun, Sang Cuijiang, Chen Jun, Yang Sibo, Liu Shihui, Du Bei, Li Jie, Yin Jianchun, Xia Xianfang, Hou Weihua, Lin Shaoping, Sun Xuesong, Zhang Hanying, Zhang Kun, Jiang Jinhui, Han Guoxing, Zheng Jianli, Xie Hengyuan, Ma Long, Huang Yin, Tian Shengli, Yu Lirong, Chen Gang, Zhao Mingyong, Min Zhaohui, Yuan Wenzhong, Cao Bin and Lin Jinming.

Contents

1 General Provisions ··· 1
2 Railway Line and Traction Power ·· 2
3 Station and Yard ·· 3
   3.1 General Requirements ·· 3
   3.2 Station ·· 3
   3.3 Depot (post), Workshop and Work Section ·· 4
4 Building Construction, Heating, Ventilation and Air Conditioning ··············· 5
   4.1 General Requirements ·· 5
   4.2 Building Envelope ·· 5
   4.3 Cold and Heat Sources for Air Conditioning and Heating System ··········· 7
   4.4 Heating ·· 7
   4.5 Ventilation and Air Conditioning ·· 8
   4.6 Testing, Controlling and Metering ··· 9
   4.7 Building Water Supply and Drainage ··· 9
5 Locomotive Facilities, Rolling Stock Facilities, EMU Facilities, Machinery and Power
   Equipment ·· 10
   5.1 General Requirements ·· 10
   5.2 Locomotive Facilities, Rolling Stock Facilities and EMU Facilities ········ 11
   5.3 Machinery and Power Equipment ··· 11
6 Electrification ··· 12
7 Electric Power ·· 13
   7.1 Power Supply and Distribution System and Equipment ························ 13
   7.2 Substation and Distribution Station ·· 14
   7.3 Power and Lighting ·· 14
8 Water Supply and Drainage ··· 16
   8.1 General Requirements ·· 16
   8.2 Water Supply ··· 16
   8.3 Drainage ·· 16
9 Communication, Information and Signaling ·· 18
   9.1 General Requirements ·· 18
   9.2 Communication and Information ··· 18
   9.3 Signaling ·· 18
Words Used for Different Degrees of Strictness ··· 20

1 General Provisions

1.0.1 The Code is prepared with the view to implementing *Energy Conservation Law of the People's Republic of China* and relevant laws and regulations as well as national guidelines and policies concerning energy conservation, to unifying the standards for design of energy conservation for construction of railway projects and to rationally allocating and efficiently using energy and resources in the construction of railway engineering.

1.0.2 This Code is applicable to the design for new railway or upgrading of existing railway.

1.0.3 For the railway construction projects, the layout of railway facilities shall be planned in coordination with local conditions and in compliance with the requirements on the development of railway transport and energy conservation, so as to provide conditions for operation, management and maintenance.

1.0.4 Railway construction projects shall use new technologies, new processes, new materials and new equipment that are featured with high efficiency, low energy consumption and capable of comprehensive utilizing resources. Production processes and equipment prohibited by the government must not be used.

1.0.5 The design of energy conservation for railway engineering shall comply with the requirements specified in the energy conservation assessment report and the approval review report, and the energy efficiency index of equipment shall not be lower than the energy conservation assessment value.

1.0.6 For railway construction projects, new energies such as solar energy, geothermal energy and wind energy, and renewable energies shall be utilized comprehensively according to the energy policies and resource conditions in the project area, so as to improve the utilization efficiency of energy and resource.

1.0.7 The energy transmission (distribution) systems for railway construction projects shall be equipped with metering devices according to the requirements in *General Principles for Equipping and Managing of the Measuring Instrument of Energy in Organization of Energy Using* (GB 17167) and relevant national standards.

1.0.8 In addition to this Code, the energy conservation design of railway engineering shall also comply with relevant provisions specified in current national standards.

2 Railway Line and Traction Power

2.0.1 Route shall be selected and main technical standards shall be determined by fully considering the requirements of energy conservation.

2.0.2 The level curve radius shall be selected according to the local conditions. In case small radius is designed under difficult conditions, such curves should be set in a concentrated way.

2.0.3 In the design of longitudinal profile, long slope and small gradient difference should be used.

2.0.4 The pusher grade section shall be set in a concentrated way, and it should be adjacent to the district station or other stations with locomotive facilities.

2.0.5 In case that freight flows in heavy-loaded and light-loaded train directions are significantly unbalanced and there are remarkable economic benefits by using different ruling grades in two directions, the ruling grades may be selected respectively in two directions.

2.0.6 Tunnels should be arranged on straight track. If a tunnel has to be arranged on a curve due to the restriction by topographic and geologic conditions, such curve should be arranged near the portal and use a relatively large radius. Tunnels should not be arranged on a reverse curve.

2.0.7 For a tunnel with mechanical ventilation, the design gradient should be small if the requirements on the minimum gradient in tunnel are met.

2.0.8 CWR shall be designed for main line of new and upgraded railways.

2.0.9 Lines with conditions allowing electric locomotive as the traction power shall use electric locomotives.

2.0.10 Locomotive model shall be selected by considering local natural conditions, and locomotives with low energy consumption shall be chosen if requirements on traction mass and transport capacity are met.

2.0.11 The AC-DC-AC locomotives with regenerative braking technology should be adopted.

2.0.12 During preparation of transport plan, the traffic control department shall balance car flows in both directions, reduce the operation of underload train, less-formation train and single locomotive, organize non-stop train and avoid repeated marshalling, or angular or roundabout car flow.

3 Station and Yard

3.1 General Requirements

3.1.1 In the design, the scale of railway station/depot (post) shall be determined according to the short-term needs and the long-term development conditions.

3.1.2 Design for station/yard shall comply with the following provisions:

  1 The distances between longitudinal yards shall shorten the section without turnout if the requirements for operation and long-term transport are met.

  2 The longitudinal profile design of station approach should facilitate train arrival, stop, start and leave.

3.1.3 Technical operation station shall be arranged at the junction between trunk lines or between trunk line and branch line and shall be helpful for reducing angular operation of trains within the station.

3.1.4 In case of a large number of angular car flows at the hub, connecting lines may be built after comparison in technology and economy.

3.1.5 The layout type of the station shall be determined based on the business type, handling procedure, operation volume and terrain conditions of the station.

3.1.6 The layout of station shall provide conditions for organizing through and non-stop trains.

3.1.7 Equipment layout in technical operation station shall be in favor of reducing repeated marshalling of train and improving operation efficiency.

3.1.8 Layout of the throat area in station shall meet the station's requirement on handling capacity and shall be able to reduce the conflict between operations.

3.1.9 CWR shall be used for receiving-departure tracks in stations for high-speed railway and inter-city railway, and it should be used for receiving-departure tracks in stations for mixed traffic railway running EMU and heavy-haul railway.

3.1.10 Untwining line for station approach shall ensure smooth passing through of main car flow and a short operation distance, and the layout of the untwining area shall be compact.

3.1.11 Surface rain water and snow water within the station should be discharged naturally.

3.2 Station

3.2.1 Design of district station shall comply with the following requirements:

  1 Location and scale of the district station shall be determined according to the layout of railway network and work division of adjacent technical operation stations.

  2 Transverse type layout should be used if the operation requirements are met.

3.2.2 Design of marshalling station shall comply with the following requirements:

  1 One marshalling station should be set at a hub, and a technical operation station may be set for auxiliary operation, if necessary.

  2 For the marshalling station mainly handling transfer and remarshalling, its location shall make the path of main car flow the shortest.

3 For the marshalling station handling both network transfer and local car flow, its location shall ensure smooth path of transfer car and reduce the distance between it and the area to be served.

3.2.3 Design of freight station shall comply with the following requirements:

1 Freight station shall be selected in coordination with the overall planning for urban construction and logistics, and the location should be close to the main technical operation station of the hub, main industrial area, container collection and distribution area, and the urban logistics park.

2 Transverse type layout should be used for freight station.

3 The existing station should be used as a freight station after renovation, and the inbound and outbound car flows at all directions shall be smooth to avoid the angular car flow.

4 Freight station shall facilitate connection and coordination with other transportation modes, and shall reserve conditions for future development.

5 Functional areas of the freight station shall be arranged logically to reduce transshipment.

3.2.4 Freight yard should be close to the main freight source.

3.2.5 Design of passenger station shall comply with the following requirements:

1 Passenger station shall be arranged to reduce the angular operation and the traveling distance of passenger train.

2 Passenger station should be designed in coordination with passenger transport facilities in city, port, airport, etc.

3 Transverse type layout should be used for passenger station.

3.2.6 Design of shunting yard and the hump shall comply with the following provisions:

1 The hump should use automatic system.

2 The plane and profile of hump head and shunting yard tail shall be designed to meet the requirements on the handling capacity of train marshalling.

3 The hump shall be set along the direction of main car flow and should use terrain condition and the prevailing wind direction.

4 Speed control device and anti-coasting device of the shunting yard shall be energy-saving and free-maintenance.

3.3 Depot (post), Workshop and Work Section

3.3.1 Layouts and scale of the test center, maintenance workshop, work section and operation maintenance depots of large track maintenance machinery shall be arranged in coordination with railway network planning.

3.3.2 Locomotive depot (running shed), rolling stock depot (running shed) and EMU depot (running shed), maintenance workshop, work section and operation maintenance depots of large track maintenance machinery should be close to the stations. The transfer track for depot shall be smooth and able to reduce the traveling distance of cars and cross interference.

3.3.3 Bases for passenger car servicing and logistics supply shall be close to the passenger transport depot, train section, EMU depot (running shed) or passenger car servicing depot.

3.3.4 Test and maintenance facilities and car stabling sidings shall be arranged in coordination with the overall layout of the station and shall reduce the invalid traveling distance of cars.

4 Building Construction, Heating, Ventilation and Air Conditioning

4.1 General Requirements

4.1.1 The energy conservation design for the railway buildings shall comply with provisions specified in national and local energy conservation standards.

4.1.2 The sunlight in winter and natural ventilation in summer shall be utilized in the designs of architectural general plan, architectural general layout and architectural plane of buildings. The major orientation of building should be arranged according to the optimal orientation of the region or close to the optimal orientation, and the prevailing wind direction in winter should be avoided.

4.1.3 The layout of production buildings, office buildings and living buildings shall be compact and shall comply with the requirements on the service function and fire separation distance. Buildings in the same area with similar function shall be built together.

4.1.4 In the designs of architectural general layout and architectural plane of buildings, the locations of the cold and heat sources and fan rooms shall be determined based on the general layout, so as to reduce the conveying distance of the cold/hot water system and the air system.

4.1.5 For the building with certain environmental requirements on personnel and mechanical equipment, the thermal design shall be carried out according to the service requirements and the environmental conditions of the place where the building is located in.

4.1.6 In the stage of construction drawing design, the calculation of thermal load shall be carried out for each room of the central heating system; the calculation of thermal load in winter and the hourly thermal load in summer shall be carried out for the air-conditioned area of the air conditioning system. In the process of scheme comparison and preparation of operation strategies, the annual dynamic load shall also be calculated.

4.1.7 During scheme comparison for the HVAC system (heating, ventilation and air conditioning system), the analysis shall be carried out based on the annual energy consumption, and the simulation calculation for the annual energy consumption shall be conducted if necessary.

4.2 Building Envelope

4.2.1 The design for thermal insulation of building envelope of railway buildings shall comply with provisions specified in *Code for Thermal Design of Civil Buildings* (GB 50176), *Design Standard for Energy Efficiency of Public Buildings* (GB 50189), *Design Standard for Energy Efficiency of Residential Buildings in Severe Cold and Cold Zones* (JGJ 26), *Design Standard for Energy Efficiency of Residential Buildings in Hot Summer and Warm Winter Zone* (JGJ 75), *Design Standard for Energy Efficiency of Residential Buildings in Hot Summer and Cold Winter Zone* (JGJ 134) and local standards for energy efficiency of buildings.

4.2.2 In severe cold and cold zones, when the area of an individual public building is more than 800 m^2, the shape coefficient shall be less than or equal to 0.40; when the area is more than 300 m^2 and less than or equal to 800 m^2, the shape coefficient shall be less than or equal to 0.50; when the requirements cannot be met, judgment and adjustment shall be made according to the relevant provisions specified in

Design Standard for Energy Efficiency of Public Buildings (GB 50189).

4.2.3 High-efficient thermal insulation technology shall be used in building envelope, the non-toxic and renewable recycled materials that are featured with light weight and low water absorption should be used, and the requirements on fire safety shall be met.

4.2.4 In severe cold and cold zones, the thermal insulation measures shall be taken between the external wall of the heating house and the ground. The depth of the external insulation material of the external wall embedded into the ground shall not be less than 800 mm.

4.2.5 These structural elements shall be provided with thermal insulation layers: the overhead floor whose bottom contacts outdoor air such as arcade building or overhead building with the need of heating; the basement top slab and the ceiling of station building porch with no need of heating.

4.2.6 The place of thermal bridge of the building envelope shall be provided with thermal insulation measures.

4.2.7 The roof may be provided with the following thermal insulation measures:

  1 The overhead roofs shall be used in hot summer and warm winter zone.

  2 The slope roof shall be provided with the attic floor for ventilation with the openable window sash or funnel cap.

  3 Greening the roof.

4.2.8 The design of door and window shall comply with the following requirements:

  1 The heat transfer coefficient, shading coefficient, visible light transmittance, window-wall area ratio, openable area of external window, air tightness, setting conditions of bay window and opening mode of windows shall comply with provisions specified in national and local standards for energy conservation design of buildings.

  2 The external window air-tightness and the area ratio of the transparent part of the roof of the production building shall be determined according to the building type, nature, production process requirements and effect of energy conservation.

  3 Gaps between external door and external wall as well as between external window and external wall shall be filled with thermal insulation and sealing materials.

  4 When using external-wall thermal insulation, the wall areas around the openings of doors and windows shall be treated by thermal insulation measures.

4.2.9 The external door of public building and residential building in severe cold zones shall be provided with porch. The external door of public building and residential building in cold zones should be provided with porch or other measures to reduce cold air infiltration. The external door of building in other areas shall be provided with thermal insulation measures. For passenger stations equipped with air conditioning and heating systems, the external door at the entrance shall be provided with porch. The external door at the entrance of the ticket hall should be provided with porch.

4.2.10 Regarding buildings in hot summer and warm winter zones as well as in hot summer and cold winter zones with large cooling load, the external window or the curtain wall shall be provided with sun-shading measures, and the shading coefficient shall comply with provisions specified in national and local standards.

4.2.11 The building structure shall comply with the requirements on structure service life specified in the current standard *Unified Standard for Reliability Design of Building Structures* (GB 50068), and should be provided with manufactured components such as prefabricated members, steel structures, multi-functional composite walls, finished railings and canopies.

4.3 Cold and Heat Sources for Air Conditioning and Heating System

4.3.1 Heating shall be designed in the forms of district heating or central heating derived from combined heat and power generation. For buildings along railway without conditions of central heating, their heating may be designed in the forms of ground-source heat pumps, water-source heat pumps and air-source heat pumps according to the local energy policies and conditions.

4.3.2 The cooling and heating devices of the air conditioning and heating systems should be placed together. The device selection shall be determined according to the building scale, service requirements, local energy source conditions, and environmental protection requirements, and shall comply with the following provisions:

1 In areas with available and useable waste heat derived from thermal power plant or from other plants, the waste heat should be utilized.

2 In area with adequate natural gas supply, the CCHP technology (distributed combined cooling, heating and power technology), gas-fired boiler technology and the gas-fired air conditioning technology should be used.

3 In area with various energies (heat, electric power, gas, etc.), the composite-energy cooling and heating technology may be used.

4.3.3 The boiler selection shall be determined according to the local resource conditions through the technical and economic comparison, and shall comply with the following provisions:

1 The coincidence factor of the thermal load shall be counted during determining the boiler capacity.

2 The coal-fired boilers with decoupling combustion technology should be used.

3 The output rating, number and performance of the boiler shall be adapted to the changes in thermal load.

4 The waste heat generated from boilers shall be fully utilized.

4.3.4 The capacity and number of auxiliary machines of boilers shall be adapted to the changes in thermal load of boilers, and the performance curve of the circulating water pump shall be in line with the performance curve of pipeline.

4.3.5 The water treatment system, circulating system, water supply system and combustion system of boiler shall be provided with automatic adjustment devices. The steam boiler with single capacity of 10 t/h or above or hot-water boiler with power of 7.0 MW or above should be provided with the centralized control system.

4.3.6 The energy-consumption devices such as fans and pumps used in cold and heat sources of air conditioning/heating systems and used in transmission/distribution systems should be in accordance with the requirements of national energy efficiency index of Grade 1.

4.4 Heating

4.4.1 The central heating for civil buildings shall use the hot water as the heating medium. Heating for production buildings should use the hot water as the heating medium.

4.4.2 Control and adjustment technology for heating network shall be used in the outdoor heating network and the indoor heating system.

4.4.3 When heating is required by local operating region of production workshop, measures for local heating shall be taken.

4.4.4 High and large space of the civil building should be provided with the radiant heating.

4.4.5 Heat transfer efficiency of outdoor heating network shall not be less than 92%, and the method of branched directly-buried installation should be used in hot-water heating network.

4.4.6 The ratio of electricity consumption to transferred heat quantity of the hot-water circulating pump for central heating shall comply with provisions specified in *Design Standard for Energy Efficiency of Public Buildings* (GB 50189).

4.4.7 Frequently-opened external doors located in factory buildings and public buildings equipped with heating and air-conditioning facilities shall be provided with air curtains.

4.4.8 The thermal insulation of thermal equipment, pipe and its fittings shall comply with relevant provisions specified in *Guide for Design of Thermal Insulation of Equipment and Pipes* (GB 8175).

4.5 Ventilation and Air Conditioning

4.5.1 Waste heat, residual wet and non-hazardous substances generated in buildings should be eliminated by the organized natural ventilation. When natural ventilation cannot satisfy the sanitary standards, mechanical ventilation or local cooling facilities shall be provided.

4.5.2 When air conditioning is required by local operating region of production workshop, measures for local air conditioning shall be taken.

4.5.3 One central air conditioning system should not be used for the different areas with different needs of air-conditioning.

4.5.4 The energy consumption per unit air volume (W_s) of draft fan in the air conditioning system and the energy efficiency ratio for conveying of hot and cold water system shall comply with provisions specified in *Design Standard for Energy Efficiency of Public Buildings* (GB 50189).

4.5.5 Rooms with high density and big change in number of people such as waiting areas (rooms) of the station and ticket halls shall be provided with fresh air demand control technology.

4.5.6 The design of the air system for air conditioning shall comply with the following provisions:

    1 When the height of building space is greater than or equal to 10 m with the volume is more than 10 000 m³, the stratified air conditioning system shall be used.

    2 If there are conditions, the air supply mode of displacement ventilation should be used for the air supplying of air conditioner.

    3 The air conditioning system with function of centralized air exhausting should be provided with the heat recovery measures.

    4 During design for the fan coil plus fresh air system, the fresh air shall be directly sent to the air-conditioned areas.

    5 If there is large calorific value on the top storey of building or on the upper part of ceiling, or if the ceiling space is relatively high, the air should not be returned directly from the interior of ceiling.

    6 In the case of full air conditioning system, the fresh air shall be utilized to the maximum.

    7 The air flow distribution design shall be carried out for the air-conditioned areas.

4.5.7 The design of the water system for air conditioning shall comply with the following provisions:

    1 The closed type circulating technology for air conditioning shall be used in the chilled water circulating system, and should be used in the cooling water circulating system.

    2 The design temperature difference between supply water and return water of water chilling unit shall not be less than 5℃. If technically reliable and economically rational, the temperature difference between supply water and return water shall be increased.

3 In case the scale of the water system is small, and at the same time, the difference in water resistance among loops is not large and the load change during the operation of the system is relatively small, the primary pump system should be used. According to the technical and economic verification, the operation adjusting mode that is featured with variable speed and variable flow may be used for the primary pump.

4 In case the scale of the water system is large, and at the same time, the difference in water resistance among loops is large and the load change during the operation of the system is relatively large, the secondary pump system shall be used. According to the change in the flow demand, the adjusting mode that is featured with variable speed and variable flow should be used for the secondary pump.

5 Cooling tower shall be placed in area with good air circulation conditions, and environmental pollution caused by it shall be avoided.

4.5.8 The cold insulation of the refrigeration equipment and air conditioning pipes shall comply with relevant provisions specified in *Guide for Design of Cold Insulation of Equipment and Pipes* (GB/T 15586).

4.6 Testing, Controlling and Metering

4.6.1 The central HVAC system should be monitored and controlled. The large centralized air conditioning system should be provided with the automatic control and detection systems.

4.6.2 The boiler room, heat exchanger room and the refrigeration machine room shall be metered, and the main content is as follows:

1 Fuel consumption.
2 Power consumption of refrigeration machine.
3 Heating capacity of central heating system.
4 Amount of water supplement.

4.6.3 When regional cold/heat source is used, the cooling/heating capacity metering devices shall be installed at the inlet of the cold/heat source of each building. When the central heating and air conditioning system is used, separate cooling/heating capacity metering devices should be installed for different users or regions. The metering device should be of the intelligent type, and the interface for remote monitoring system should be reserved.

4.6.4 The heating and air conditioning system shall be equipped with the room temperature control device. The radiator and the radiant heating system shall be equipped with the automatic temperature control valves.

4.7 Building Water Supply and Drainage

4.7.1 The water instruments, pipes, fittings, valves and metering devices used for building water supply and drainage shall comply with relevant provisions specified in *Technical Conditions for Water Saving Products and General Regulation for Management* (GB/T 18870).

4.7.2 The hydraulic pressure of outdoor water supply network shall be utilized for building water supply.

4.7.3 The heat source of the hot water supply system shall be determined according to the service requirement, heat consumption, distribution of water consumers and heat source conditions, and the excess heat, waste heat, renewable energy or air-source heat pump should be used as the heat source.

4.7.4 The railway buildings shall be equipped with reclaimed water and rainwater utilization facilities according to relevant technical policies and local conditions.

5 Locomotive Facilities, Rolling Stock Facilities, EMU Facilities, Machinery and Power Equipment

5.1 General Requirements

5.1.1 The long route and crew shift system shall be used for the operation of the locomotive, and the layout and scale of locomotive, rolling stock and EMU (electric multiple unit) facilities shall be determined rationally. Specialized and centralized repair should be used for the maintenance of the locomotive, rolling stock and EMU.

5.1.2 The arrangement of locomotive, rolling stock and EMU depots (running sheds) and operation maintenance depots of large track maintenance machinery shall comply with the following provisions:

    1 The transport distance of the repair parts, materials and spare parts shall be reduced in general layout plan.

    2 Buildings shall be designed in the forms of combined factory buildings, multi-storey buildings or comprehensive multi-storey buildings.

    3 The powerhouse shall be located close to the load center.

5.1.3 Process pipes shall be laid integratedly, and all kinds of pipelines shall be short and straight.

5.1.4 For all kinds of specialized facilities, the motor power and number shall be allocated rationally. In the case of electric traction equipment with large capacity and frequent starting operation, the energy saving technologies such as frequency control of motor speed shall be used.

5.1.5 Electric heating and drying equipment shall be provided with technologies such as far-infrared heating and induction heating, and the electric energy utilization efficiency shall comply with relevant provisions specified in *Monitoring and Testing Method for Energy Saving of Electroheat Device in Industry* (GB/T 15911).

5.1.6 The locomotive, rolling stock and EMU depots (running sheds), tank washing point and operation maintenance depot of large track maintenance machinery shall be equipped with waste oil recovery devices.

5.1.7 Production process water generated from washing and cooling devices for outer cover and spare parts of locomotive, rolling stock, EMU and large track maintenance machinery shall be recycled.

5.1.8 The design for compressed air station shall comply with the following provisions:

    1 The compressed air station shall be located close to the air load center, and the intakeport of the compressor shall be installed in a shaded place free from heat source.

    2 Air compressor units shall be in accordance with the product requirements on high efficiency and low energy consumption.

    3 The air compressor should be provided with the air cooling mode. If the water cooling mode is used, the recycle rate of water shall be more than 90% and the waste heat shall be recycled.

    4 The compressed air shall be conveyed at an economic flow speed.

    5 The type selection of the air compressor shall comply with provisions specified in *Minimum Allowable Values of Energy Efficiency and Energy Efficiency Grades for Displacement Air Compressors* (GB 19153).

    6 Metering devices shall be set separately.

5.1.9 If the requirement on the fire separation distance is met, the site of oil storage should be located close to the locomotive servicing depot. The gravity flow should be used for oiling the aboveground oil storage with the help of the natural conditions.

5.2 Locomotive Facilities, Rolling Stock Facilities and EMU Facilities

5.2.1 The oil storage tanks and pipes of the locomotive depots (running sheds) in the severe cold zone and the cold zone shall be provided with thermal insulation measures.

5.2.2 The diesel locomotive depots in the severe cold zone and the cold zone should be equipped with the locomotive temporary rest sheds or the surface preheating and thermal insulation devices.

5.2.3 For the locomotive depots (running sheds), the sunlight should be utilized to dry sand. The sand drying and conveying equipment shall be the energy-saving equipment.

5.2.4 The following depots (posts) and workshops should be built together:

1 The passenger car technical servicing depot or the freight car repair yard at station and the rolling stock depot should be built together when they are in the same station.

2 When the freight car repair yard at station is adjacent to the train inspection yard, the compressed air stations should be built together.

3 The auxiliary repair workshop and the depot repair workshop for mechanical refrigerator cars should be built together.

4 In the locomotive depots (running sheds) or EMU depots (running sheds), the house used for wheelset tread diagnosis equipment and the house used for dynamic testing equipment of electric pantograph should be built together.

5.3 Machinery and Power Equipment

5.3.1 The quarry shall be equipped with the crushing, screening, washing and transport equipment according to the specifications of ballast and the following provisions:

1 The landform such as mountain land and slope shall be utilized for the process layout of the quarry. The raw stone silo for the use of the stone crusher should be located close to the quarrying area.

2 The by-products shall be utilized in the process of quarrying, and the tailings shall be utilized comprehensively as per classifications.

3 Water for washing ballast shall be recycled.

5.3.2 Elevator and escalator shall be designed in accordance with the requirements of energy-saving products with frequency conversion device.

5.3.3 Electric drive mode should be used in the loading and unloading machinery. The voltage level of the power supply shall be determined comprehensively based on the installed power of equipment, power supply conditions, line loss, transformer loss, etc. The power-driven loading and unloading machinery shall be equipped with energy consumption metering devices.

5.3.4 The conveyor and crane for coals and ores should be driven by variable frequency motors, taking into consideration cost efficiency.

6 Electrification

6.0.1 In the design for the traction power supply system, the power-saving technologies promoted by the state shall be employed. The power supply solution shall be designed to reduce the power transmission loss and to improve the efficiency of traction power supply.

6.0.2 The voltage level of incoming power supply of the traction substation for high-speed railways shall be 220 kV and above. The voltage level of incoming power supply of the traction substation for heavy-haul railways should be 220 kV and above. The voltage level for other railways should be 110 kV and above.

6.0.3 When the power factor of the traction power supply system is to be improved, reactive power compensation devices shall be installed in the traction substation (or switching post or sectioning post). After compensation, the average power factor at the primary side of the traction substation shall not be less than 0.9.

6.0.4 The AT feeding mode of 2×25 kV should be employed for the contact line system of the main line for the high-speed railway and the heavy-haul railway. The direct feeding mode of 1×25 kV with return conductor may be employed for connecting lines in hub areas, running tracks for EMU and EMU depots (running sheds and yards). The direct feeding mode of 1×25 kV with return conductor or the AT feeding mode of 2×25 kV may be employed for the contact line system of other railways.

6.0.5 The traction substation shall be close to the load center.

6.0.6 The installed capacity of the traction transformer shall be determined according to the short-term traffic volumes, and the overload capability of transformer shall be fully utilized.

6.0.7 The OCS of the double-track railway or the multiple-track railway may be provided with the parallel feeding at the end. The OCS of the high-speed railway or the heavy-haul railway may be provided with the entirely parallel feeding mode. For the roundabout route section, bypass feeders shall be used for OCS.

6.0.8 Traction transformers shall be energy-saving transformers featuring with high capacity utilization ratio, better overload capability and low loss.

6.0.9 Energy-saving traction power supply equipment shall be used.

7 Electric Power

7.1 Power Supply and Distribution System and Equipment

7.1.1 The design of the power supply for the power supply and distribution system shall comply with the following provisions:

1 The power supply of 10 kV and above from public grid should be used for power supply and distribution. If it is unavailable, the power supply at other voltage levels from public grid may be used. In case of areas with no power supply from public grid, photovoltaic energy, wind power or other available clean energies should be used for power supply. For electrified sections, the power supply from the traction power supply system may be used.

2 For railway passenger stations and production and auxiliary buildings, the solar photovoltaic system and the trigeneration or combined cooling, heating and power(CCHP) system may be used as the auxiliary power supply according to the scale of the buildings and the regional conditions.

7.1.2 The design of the feeding mode shall comply with the following provisions:

1 The power transforming and distributing equipment shall be close to the main load center.

2 The economical operation mode shall be used so as to reduce the links of intermediate power transforming.

3 Power supply systems for production and living shall be separated.

7.1.3 The selection of conductors of 35 kV and above shall be verified according to the economical current density.

7.1.4 The remote control technology shall be used for the power supply network. If the condition does not permit, technical provisions for remote control shall be made.

7.1.5 In the process of selection of transformer capacity, line layout and laying method, the line impedance shall be reduced and the natural power factor shall be improved.

7.1.6 Power supply and distribution system with natural power factor not meeting the specified requirement shall be equipped with the reactive power compensation device and shall comply with the following provisions:

1 The power factors of the HV and MV systems shall be over 0.9 and the power factor of the LV system shall be over 0.85.

2 The capacity and phase of the compensation device shall be set according to factors such as compensation amount, facilitating voltage regulating and no resonance.

3 The single equipment having significant effect on the power factor of the power supply and distribution system and operating continuously may be provided with the local compensation. For system with power factor fluctuating greatly, the reactive power compensation device may be used.

4 Power lines mainly using cables shall be equipped with inductive reactive power compensation devices according to the length, material, specifications, load conditions and laying conditions of the cable.

5 The reactive power compensation device shall have the function of preventing the reverse reactive energy to the grid. If the power supply and distribution system has severe harmonic current, the reactive power compensation device shall have the filtering function.

7.1.7 The selection of the power transformer shall comply with the following provisions:

1 The three-phase distribution transformers and the power transformers shall comply with provisions specified in *Minimum Allowable Values of Energy Efficiency and Energy Efficiency Grades for Three-phase Distribution Transformers* (GB 20052) and *Minimum Allowable Values of Energy Efficiency and Energy Efficiency Grades for Power Transformers* (GB 24790).

2 The operation of transformers shall comply with provisions specified in *Economical Operation for Power Transformers* (GB/T 13462).

3 If the load of main transformer has significant variations in day and night and in different seasons, small capacity transformers in line with the regular variation amount may be additionally installed and the light-load switching devices should be used.

7.1.8 The design of the railway energy management system shall comply with the provisions specified in *Standard for Design of Intelligent Building* (GB/T 50314).

7.2 Substation and Distribution Station

7.2.1 The primary equipment of the substation and distribution station should be the minimum oil or the oil-free equipment. The automation technologies shall be used for protection, control and management.

7.2.2 For the design of electrical power supply, technologies such as time-of-use pricing and power load control may be adopted according to the local charging rules or requirements of power supply departments.

7.3 Power and Lighting

7.3.1 The power supply for power and lighting of the LV power distribution system shall be separated.

7.3.2 When the LV power distribution system connects to the AC220 V or AC380 V single-phase (two-phase) consumers, the three-phase load should be balanced.

7.3.3 The power distribution and control of the welding machine shall comply with the following provisions:

1 Each welding machine shall be controlled separately.

2 Multiple single-phase welding machines should be uniformly located in the three-phase line.

3 Small and medium welding machines frequently subject to no-load operation may be equipped with the no-load automatic stop devices.

7.3.4 For the AC motor with requirement on speed control as well as fans or pumps with requirements on change of air or water volume, the control element shall have the variable frequency drive(VFD).

7.3.5 For the electrical lighting, high-efficacy energy-saving luminaires complying with the national energy efficiency standards shall be employed, and the lighting design shall comply with the provisions specified in *Standard for Lighting Design of Buildings* (GB 50034) and other relevant national standards.

7.3.6 For the lighting design, the types of general lighting, localized lighting, local lighting and group control shall be determined according to the function of space and illuminance standards.

7.3.7 Except for special needs, the electric light source shall be the product with national energy-saving certification, and shall meet the requirements on the minimum allowable values of energy efficiency and energy efficiency grades stipulated by the state.

7.3.8 The daylight shall be fully utilized for the lighting design. For interior parts of buildings and structures, daylight redirecting devices and reflective devices may be used to introduce the daylight into the interior parts for lighting.

7.3.9 For outdoor lighting, complete sets of photovoltaic and wind-power lighting devices may be adopted according to the natural environmental conditions.

7.3.10 Gaseous discharge lamps with inductive ballasts shall be provided with capacitance compensation. The power factor of the fluorescent lamp shall not be less than 0.9 and the power factor of the high intensity discharge lamp shall not be less than 0.85.

7.3.11 The lighting system shall be provided with control devices based on decentralized, centralized, manual and automatic control as well as requirements on characteristics, functions, standards and application of lighting, and shall comply with the following provisions:

    1 Station buildings of large and super large passenger stations as well as other large buildings with complex functions and higher requirements on lighting environment shall be provided with the intelligent lighting control system.

    2 The manual control, sunlight control, time control or the combined control mode shall be adopted for the outdoor lighting.

    3 For landscape lighting, controls with different modes and different scenarios shall be provided.

7.3.12 Power consumption metering shall comply with the following provisions:

    1 The production and living power consumptions shall be metered separately.

    2 The metering shall be based on different users.

    3 Consumers with large energy consumptions shall be metered separately.

    4 In addition to the metering devices for charging, the low-voltage power distribution system shall also be provided with metering devices for assessment as per the equipment category, power consuming area, divisions or functional areas.

    5 The intelligent metering devices and remote monitoring systems should be used.

8 Water Supply and Drainage

8.1 General Requirements

8.1.1 In the design for water supply and drainage engineering, energy saving shall be regarded as one of the main factors for scheme comparison, system optimization and equipment selection.

8.1.2 The equipment, pipes, fittings and valves used for water supply and drainage engineering shall meet the current national technical requirements on energy-saving products.

8.1.3 In production, water should be recycled, reclaimed and used efficiently.

8.2 Water Supply

8.2.1 In the design for water supply engineering, the separate water supply system, dual water supply system or partial pressure water supply system shall be used according to users' requirements on water quality, water volume and water pressure. For water diversion and water conveyance works, the scheme of water conveyance by gravity flow or by local pressurization should be used.

8.2.2 Railway water supply should be derived from urban water supply system. In water resource shortage areas, brackish water or sea water and rainwater may be utilized.

8.2.3 The direct water supply, pressure-superposed water supply and variable frequency water supply should be used for the water supply station. The direct water supply and water supply with high level water tank and water tower should be used for the domestic water supply station/point. For the secondary pressurized water supply, inlet pipes should be arranged in parallel to connect with the urban tap water network so as to utilize the pressure of pipe network for water supply.

8.2.4 For layout of the water supply network, the water consumption unit subject to independent accounting shall be provided with the main water meter, and the workshops and water consumers shall be provided with the individual water meters. The installation location of water meter shall be able to facilitate reading and maintenance.

8.2.5 Water supply pipes shall be made from clean and non-toxic pipe materials that are featured with small friction coefficient and high strength.

8.2.6 Leakage detectors shall be provided for management and maintenance of water supply network.

8.3 Drainage

8.3.1 For the layout of the drainage network, the terrain conditions shall be utilized for reducing the pumping times.

8.3.2 Waste water should be discharged into the urban drainage system.

8.3.3 The design for waste water treatment shall comply with the following provisions:

    1 Waste water in the same station area or region should be treated in a centralized way.

    2 In the elevation design, the terrain conditions should be utilized and the layout shall be compact.

    3 The scale of the waste water treatment plant (station) shall be determined according to the operating shift system.

4 The small and medium railway stations should be provided with the energy-efficient waste water treatment process.

5 The waste water subject to the advanced treatment should be recycled.

6 The waste water drainage outlet shall be equipped with metering devices.

8.3.4 The drainage system shall be designed according to the principle that rainwater and waste water are drained separately. In case of any demand for rainwater utilization, rainwater from large roofs and squares where rainwater pollution is relatively slight shall be collected for utilization.

9 Communication, Information and Signaling

9.1 General Requirements

9.1.1 Communication, information and signaling equipment shall be of economical, efficient and integrated products, and shall comply with the relevant provisions of the national energy-efficient electronic products.

9.1.2 The layout of communication, information and signaling equipment shall be compact, and all kinds of wiring routes shall be simple and direct. If conditions permit, buildings for communication equipment and buildings for information equipment and machinery may be combined.

9.1.3 Buildings for communication, information and signaling equipment of the station should be built in a concentrated manner. Buildings for the section signaling relay station and the signaling equipment of block post and buildings for the communication base stations may be combined.

9.2 Communication and Information

9.2.1 The indexes of power consumption and heat dissipation as well as conversion efficiency and power factor of power device shall be regarded as the necessary factors for selection of the communication and information equipment.

9.2.2 The selection of the power device shall comply with the following provisions:
　　1 The power device shall be equipped with the centralized monitoring communication interface.
　　2 The rectifying equipment shall be of the high-frequency switch equipment.
　　3 The power factor and efficiency of the power device shall comply with relevant provisions specified in the railway industry standards.
　　4 Batteries shall be of the maintenance-free type.
　　5 When the external AC power supply is unavailable or unreliable, the solar energy, wind power and other energy sources may be used for power supply.

9.2.3 The sleep mode should be provided for the high-frequency switching power supply device and UPS equipment.

9.2.4 Communication and information lines should be made up of optical cables.

9.3 Signaling

9.3.1 In the process of circuit design, if the requirements on safety and reliability are met, the relay energized in normal status should be of the high-resistance type.

9.3.2 The capacity of the power supply equipment shall be determined via calculation. If the requirement on power utilization of signaling equipment is met and the development conditions are reserved, the category and number of power supply panels shall be reduced. The comprehensively intelligent power supply equipment with high power factor should be provided for railway sections and stations.

9.3.3 The mode of indoor AC output and local rectification or variable voltage power supply in outdoors should be used for outdoor low-voltage power supply.

9.3.4 The computer interlocking mode should be used for the station interlocking equipment. The hump control equipment should be provided with the automatic control system.

9.3.5 The optical cables should be used for the information transmission between stations.

9.3.6 The mode of distributed control via separated signal line and separated operation shall be used for the No. 18 turnouts and those with numbers greater than 18.

9.3.7 In the process of calculating the heat dissipation of the signaling equipment, the power consumption of the instantaneous operating equipment such as the switch equipment shall be deducted and the heat dissipation coefficient of individual equipment shall be selected.

Words Used for Different Degrees of Strictness

In order to mark the differences in executing the requirements in this Code, words used for different degrees of strictness are explained as follows:

(1) Words denoting a very strict or mandatory requirement:

"Must" is used for affirmation; "must not" is used for negation.

(2) Words denoting a strict requirement under normal conditions:

"Shall" is used for affirmation; "shall not" is used for negation.

(3) Words denoting a permission of a slight choice or an indication of the most suitable choice when conditions permit:

"Should" is used for affirmation; "should not" is used for negation.

(4) "May" is used to express the option available, sometimes with the conditional permit.